Darting through the dim night be mistaken for birds or giant **mammals**. They have furry b produce milk to feed their young, just as cats, dogs, and humans do. Unlike other mammals, however, bats have the ability to fly!

These bats are leaving their cave at dusk to hunt insects. Most bats are active at night, when the competition for food and the danger of being attacked are less than during the day.

There are about 850 **species**, or kinds, of bats. Roughly one-fifth of these belong to a group called **megabats,** most of which are known as **flying foxes**. These bats eat fruit and are found only in warm, tropical regions.

The **Gigantic Flying Fox**, found in India, is the largest known bat. It weighs over two pounds and has a wingspan of six feet. This fruit-eater has the typical foxlike face that gave this group of bats its name.

All other bats belong to a group called **microbats**. These smaller bats eat a wider variety of foods and live throughout the world, except in the polar regions. All of the bats found in the United States (about 45 species) are microbats.

Most microbats, including the one in this photo, eat huge numbers of insects every night. By helping to control insect populations, these bats help to maintain the balance of nature.

Found in the deserts of Arizona and New Mexico, the **Lesser Long-nosed Bat** feeds on fruit, nectar, and pollen. These bats are important to their habitat because they help to pollinate plants.

The **Hoary Bat** lives in forests throughout North and South America. Its long, thick fur keeps it warm and enables it to fly in cold weather.

Shown here hopping toward its prey, the **Vampire Bat** feeds on the blood of animals. It makes a painless cut in the animal with its razor-sharp teeth and laps up a small amount of blood. Vampire bats are found only in Central and South America.

Most bats spend the night flying around in search of food. They fly by flapping their wings, which are thin layers of skin stretched across the bones of the hands and arms. Notice how long the bat's fingers are! Like the spokes of an umbrella, they fold up and tuck away the wing membrane as the wing closes.

This bat is shown in three positions as it flies. The bat flaps its wings with a downward stroke, raises and extends them, and—whoosh!—the wings come down again. This cycle of wing-flapping keeps the bat aloft and moving forward.

The relatively narrow wings of the **Mexican Free-tailed Bat** make it a swift flier. It can travel long distances each night in search of insects.

The broader wings of the **Lesser Long-nosed Bat** enable it to hover near fruit or flowers. Its wings are also suited to making sharp turns in tight spaces.

Most bats spend the day hanging head-down from a **roost**. The roost provides a place for bats to rest, groom themselves, communicate, mate, and hibernate. Depending on the species and where they live, bats may roost in trees, caves, or human-made structures such as buildings or mines.

This fruit bat roosts by hanging in a tree. Notice how its toes grasp the branch. It has wrapped its wings around its body to conserve body heat.

Here, **Big Brown Bats** are taking shelter under a wooden bridge. Found throughout North America, these bats also roost in buildings, tunnels, caves, and hollow trees.

These **Honduran White Bats** are roosting in a tent made out of a large leaf. They cling to the tent with their toes, which they slip through tiny holes they have chewed near the leaf stem.

Bats often roost huddled together in huge groups. This cave houses a nursery colony of **Mexican Free-tailed Bats**, mostly females and their young.

All bats can see. But to function in the dark, most bats rely on their keen senses of smell and hearing.

Microbats find food and get around in the dark by **echolocating**, or sending out sounds and listening for echoes. The sounds (too high-pitched for humans to hear) bounce off prey and return to the bat's ears as echoes. The bat interprets the echoes to locate the prey.

Most flying foxes use their excellent vision and sharp sense of smell to locate fruit and flowers. They also have a special layer in their eyes that reflects light and helps them see in dim light. This reflective layer causes their eyes to glow eerily when light shines on them.

The **Eastern Pipistrelle Bat** does not see as well as a flying fox. It "sees" by echolocating.

The **Tome's Long-eared Bat** sends out sounds through its nose, directing them with a flap of skin called a **nose leaf**. This bat's large, funnel-shaped ears can detect even the faintest echoes.

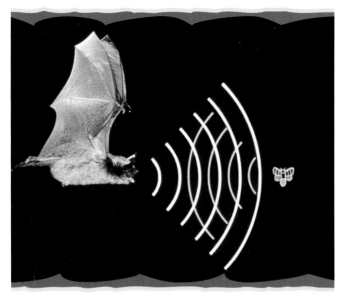

This illustration shows a bat echolocating. The white arcs stand for the sound waves sent out by the bat. The yellow arcs stand for the sound waves that have bounced off the insect, echoing back into the bat's ears.

This **Big Brown Bat** is closing in on a moth it has found using echolocation. Bats also use echolocation to avoid obstacles.

Most bats eat insects. The average microbat eats more than half its body weight in insects every night! If it weren't for bats, some insect populations might grow large enough to harm the environment.

Some microbats eat animals other than insects, such as mice, fish, frogs, or smaller bats. Vampires are the only bats that feed on animal blood.

The **Pallid Bat** eats grasshoppers, crickets, and beetles that it catches on the ground. The bat in this photo is taking its prey to a night perch to eat it there. A native of the desert, the Pallid Bat can get all the water it needs from the insects it eats.

All megabats and a few species of microbats feed on fruit, nectar, and pollen. Fruit-eating bats help the environment by scattering fruit seeds, which helps to plant trees in new places. Bats that eat nectar or pollen also **pollinate** plants, making it possible for them to produce fruit containing fertile seeds.

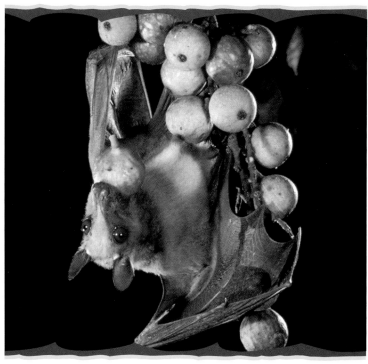

This fruit bat is eating a fig. It cannot digest the seeds, which fall to the ground in the bat's droppings. Some of the seeds sprout and become new fig trees.

This bat is about to feed on a flower. It thrusts its head deep into the blossom to lap up the nectar, getting pollen on its face. The pollen may rub off on the next flower the bat visits, pollinating the plant.

For bats, the basic family unit is a mother and her growing young. In general, male bats do not take part in family life. However, in some species, one male may roost with a group of females and their pups.

The **Gambian Epauletted Fruit Bat** usually has one pup twice a year. This bat carries her young pup with her wherever she goes, as do most megabat mothers.

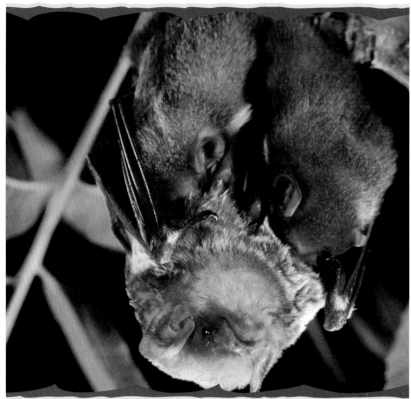

Red Bats usually have more than one pup per litter. The mother in this photo is nursing twins.

Female **Mexican Free-tailed Bats** leave their babies in cave "nurseries." Each mother returns several times a day to feed her baby. She uses her senses of hearing and smell to find her own pup among the thousands of babies in the nursery colony.

This group is what is known as a **camp** of flying foxes. One benefit of roosting together is that camp members show each other the best food sources within their area.

Food is scarce during the cold season, so many bats **hibernate**, or go into a deep sleep. This helps them save energy until food is available again in the spring.

Some bats travel to warmer places during these months. This is called **migration**.

These bats are hibernating together in a cave. Huddling close to one another helps them to conserve body heat.

Water droplets have formed on this hibernating bat because its body temperature has dropped, causing the water vapor surrounding the bat to change to liquid.

Thousands of female and young **Mexican Free-tailed Bats** are migrating from their cave in Texas to a winter roosting site in Mexico. Most of the adult males remain in Mexico all year round.

Many bat species are in danger of becoming **extinct,** or dying out, as a result of human actions. People awaken hibernating bats, causing them to use up their energy before the winter is over and starve to death. We also destroy bats' habitats, leaving them no place to live. Many people are working to stop these practices so that bats have a better chance of survival.

The **Gray Bat** is in danger of becoming extinct. Since so many of these bats hibernate together in one cave, thousands can die at one time if they are disturbed.

This **Marianas Flying Fox** is one of about 500 left. The **Lesser Marianas Flying Fox** species is already extinct.

The number of these **Lesser Long-nosed Bats** has declined sharply in recent years. One of the main causes is that their caves have been disturbed or destroyed by people.

One way in which people are helping bats is by making sure that their roosting sites are protected. For example, signs are posted outside caves to keep people from entering.

Perhaps because most bats are active at night and are difficult to observe, they are greatly misunderstood. We now know that these shy, gentle creatures play a vital role in maintaining the balance of nature. As our understanding of bats grows, so does our ability to protect them and, in turn, to protect our shared home.